MEDICAL MARIJUANA CAREGIVER'S JOURNAL

Annie –
Good Luck & Happy Trails !
Paul

MEDICAL MARIJUANA CAREGIVER'S JOURNAL

by
Chemo Sabe, M.A.

THANK YOU

Thanks to the lawyers and staff at Pier 5 Law Offices, in San Francisco. They are caring people who have become dear friends. Their office has been an oasis, their guidance indispensable, their friendship a blessing. I love them all. Hopefully, there will one day come an end to this mindless war on the sick, when caregivers will no longer need legal advisors.

DEDICATION

Dedicated to the memory of Tod Mikuriya, M.D.

Thank you, Doctor Tod! Good Luck and Happy Trails!

AUTHOR'S DISCLAIMER

CONTENTS

INTRODUCTION

Marijuana is a benign plant which has been part of the human pharmacology for 5000 years, in virtually every culture and time period, pre-dating and throughout our recorded history: from ancient times, it has been medicine for the Assyrians, Persians, Greeks, Romans and East Indians; for five millennia marijuana has been used in Chinese medicine; during the 1800s, European and American medical journals published more than 100 papers on the medical uses of marijuana; today it has been legalized in a dozen states, and it is being studied and used in some of the world's best hospitals.

Yet, at the time of this writing, our federal government holds the position that marijuana has zero medical or scientific value and refuses to declassify it as a schedule one substance more dangerous than narcotics, cocaine, amphetamines or barbiturates, even though not a single death has ever been recorded as a result of marijuana overdose, as long as the federal government has been keeping records.

The State of California, on the other hand, holds a more enlightened view. In California, patients have a right to use marijuana, with their doctor's support. Moreover, the *California Health and Safety Code* has empowered the University of California, one of the best research hospitals in the world, to study medical marijuana. In addition, the *Code* spells out procedures for doctors and researchers to apply to the state for quantities of marijuana seized in drug raids for use in medical studies. **The first study has already been done, at University of California San Francisco, and is published in the journal *Neurology*, proving the medical efficacy of marijuana. More studies are under way.**

According to Doctor Lester Grinspoon, emeritus professor of psychiatry, Harvard Medical School, "Marijuana is effective at relieving nausea and vomiting, spasticity, appetite loss, certain types of pain, and other debilitating symptoms. And it is extraordinarily safe, safer than most medicines prescribed every day. **If marijuana were a new discovery rather than a well-known substance carrying cultural and political baggage, it would be hailed as a wonder drug.**"

Doctors in the University of California San Francisco Hospital report that cancer patients have been allowed to use marijuana in their rooms for more than 30 years. This is not because it has no medical value. It is because UCSF doctors, for three decades, have seen that marijuana helps their patients.

Today, many oncologists routinely prescribe *Marinol*, for their chemo patients. *Marinol* is synthetic marijuana in a sesame oil gelatin capsule. *Marinol* can be purchased at any pharmacy. Medi-Cal, Medi-Care and other forms of insurance will pay for *Marinol*.

Doctor Tod Mikuriya is a Berkeley physician with half a lifetime of experience with marijuana as medicine. In his book, *Marijuana Medical Handbook: A Guide to Theraputic Use*, he discusses the use of marijuana to treat a surprising range of conditions, including clinical depression. According to Doctor Mikuriya, **"The power of cannabis to fight depression is perhaps its most important property. In accordance with the perverse logic of DEA bureaucrats, 'euphoria' has been listed as an adverse reaction of marijuana."**

Marijuana can keep up the spirits of those struggling with chronic or terminal conditions. It has a powerful ability to help patients keep their mood elevated and avoid the depression and negative mindset that can accompany chronic or life-threatening conditions. And without a positive mindset in the patient, no doctor or therapy will be helpful.

In my direct experience with marijuana, as a professional caregiver to cancer/chemo and hospice patients for more than a decade, I observe it to have profound ability to mitigate the side effects of chemotherapy and radiation therapy.

I've seen marijuana bring relief for a variety of serious medical conditions, including but not limited to arthritis, fibromyalgia, insomnia, anorexia, multiple sclerosis, epilepsy, glaucoma, asthma and clinical depression.

One patient controls epilepsy with marijuana alone and is able to do without more toxic drugs and their side effects. She learned this when her Medi-Cal benefits expired and she could not afford the pharmacy bill. The only medicine she was taking was the marijuana that I provided at no cost. She was fine, until she stopped the marijuana.

She was a Kaiser patient, due for a test that she feared would expose her marijuana use and bump her to the bottom of the list for a liver transplant. So she stopped the marijuana. And immediately a grand mall seizure struck. From that she learned that marijuana can control her seizures. She prefers it to the pharmaceuticals.

One 67 year-old patient with advanced prostate cancer was taking so much morphine that he was groggy and dopey all the time. And it no longer masked his pain. When he began marijuana therapy, he completely put aside the narcotic painkillers. He was able to live pain free and was no longer groggy. He had his personality and his life again. His wife took my face in her hands and, looking into my eyes, said "Thank you! My husband is back!"

It seems clear that marijuana has already become a part of our modern pharmacology. But the feds are ignoring the news. At the time of this writing, they are still arresting and prosecuting patients, caregivers and physicians, destroying medical gardens and lives.

Doctor Marian Molly Fry and her husband, lawyer Dale Schafer, had their offices raided and are currently under indictment by the federal government for serving medical marijuana patients. **"When I became a doctor, I took an oath to do what was best for my patients—not an oath of allegiance to the government or politics,"** says doctor Fry. God bless her!

The stubborn head-in-the-sand attitude of the federal government is untenable and even hypocritical: while publicly denying the medical value of marijuana, the government grows and provides (inferior quality) marijuana to at least six patients who have proven their need. Moreover, the government has permitted one pharmaceutical company to patent, manufacture and market marijuana under the name *Marinol*, formally recognizing its medical value. And in 1999 the government's own Institute of Medicine published a report, "Marijuana and Medicine: Assessing the Science Base." In part, it concluded "The accumulated data indicate a potential theraputic value for cannabinoid drugs, particularly for symptoms such as pain relief, control of nausea and vomiting, appetite stimulation (and therefore) would be moderately well suited for chemotherapy induced nausea and vomiting, and AIDS wasting."

There is clearly medical value in marijuana. It should be studied and made a useful addition to our medicine chest. What sense does it make to hide it, ignore it, be afraid of it? The wise will study, learn about and make use of it.

While the debate continues, I've been growing and providing medical marijuana at no cost to cancer/chemo patients, hospice patients and other of the sick, homeless and hopeless in our community. The genie is out of the bottle. Patients and their loved ones, doctors and researchers, health care workers and caregivers have seen the magic. Marijuana is medicine. Indeed, it is a versatile and effective medicine. It will soon become accepted. Now that so many have seen the benefits, it will be harder for the feds to put the stopper back in the bottle. But they'll keep trying. They'll keep intimidating and harming doctors, arresting patients and caregivers, destroying medical gardens and disrupting lives, until you make it clear at the voting booth that you don't want tax dollars to be wasted hurting doctors, patients and caregivers.

Some of the patients I serve have been arrested by highway patrol, local law enforcement or federal agents for using or growing medical marijuana. Patients and caregivers feel the pressure constantly. Those who are paying attention live with fear and paranoia. Such an atmosphere is not conducive to health and healing.

Perhaps the UCSF study, and others to come, and the report from the government's own Institute of Medicine will give lawmakers pause and encourage them to re-think the government position on medical marijuana.

Meanwhile, you can help patients, caregivers and doctors. Vote out the prosecutors and sheriffs who are cooperating with the feds, particularly in the Sierra and central valley. Vote out the judges who sign the warrants. Support medical marijuana initiatives. Support local and congressional leaders who support medical marijuana. And watch their performance. Some prosecutors and sheriffs have solicited the medical marijuana vote, only to take office and begin a ruthless campaign against the sick and suffering. Watch their performance.

The aim of this book is not to enter the debate, but to inform by presenting a picture of the work of one caregiver, by reporting the things I've seen and experienced during many years of work with marijuana as medicine. I've seen and learned things the public is unaware of and ought to know.

This book is organized as an anecdotal collection, little glimpses of my experiences with patients and some of the different kinds of people, conditions and situations I've encountered. There is wide variety in the people I've served and the health conditions they suffer.   They are from every socioeconomic group. They live everywhere, from the park to Park Place, the city streets to City Hall. They represent every kind of cancer and other health condition imaginable.

Cancer comes to all. It will come to you or someone close to you. Marijuana would make a difference.   Will marijuana be among your options?   Will there be a caregiver to serve you or your loved one? Will there be a co-op to supply your meds?

Caregiving is the most interesting and rewarding work I've ever done. I hope this book will inform and inspire others to become caregivers or to contribute to the support of patients and caregivers.

Chemo Sabe
San Francisco, California
Summer 2007

THE PATIENTS

PIETER

Age: 54. Architect.   Condition: terminal melanoma.

Pieter was my first patient. He was my best friend and buddy. We were brothers. His diagnosis came in August, the day before we took our girlfriends to Yosemite for a week at Tioga Pass and Tuolumne Meadows. We liked to go up to the high country in August and July, when it's hot in the lowlands. When it's brutally hot at sea level, it's always cool and refreshing at 10,000 feet.

Saddlebag Lake is a favorite spot: campsites under the pines overlooking the lake; a little log cabin trading post, so you don't have far to go for groceries or fishing tackle. Pieter brought his sketchbook and paintbox. He liked to sit outdoors and sketch studies with watercolor, for making paintings on canvas back at the studio. Saddlebag Lake is a perfect spot for it.

I brought my ultralight spinning tackle. There's some fine trout fishing in the crystal clear lakes and streams up there. And it does not get any better than trout grilled and eaten beside the stream where they were caught.

There are trails that wander through the spectacular country, offering some of the very best views of Yosemite. By August the snowmelt has usually slowed so the streams are easy to cross. So we pitched our tents at Saddlebag Lake and were set for a great week in the high country.

I had noticed that Pieter was edgy and nervous, distracted. But I passed it off, thinking he still had his mind on the business of the past week. I did not know that the business troubling his mind was the diagnosis he'd received the day before, that he had melanoma, the most aggressive and dangerous form of cancer. He told me about it that evening, as we sat by the fire. One year and one day later, Pieter was gone from this world.

During the year that followed the diagnosis, Pieter placed his hope in modern medicine. He underwent the conventional treatments. And, when those did not work, there were experimental treatments. With each treatment, he grew thinner, paler, weaker. As with many cancer patients struggling to survive, or struggling with mortality issues, he became withdrawn. With diminished strength and stamina, he could not continue to live a normal life. He retreated to his Sausalito home, where he lived in pajamas, spending almost all his time in bed or lying on the couch, growing steadily thinner and weaker. Thankfully, his mind remained clear until the last week.

His girlfriend stayed with him to provide care. And I would visit every afternoon and evening. Pieter restricted visits from family and friends, requiring them to make appointments thru his girlfriend. He seldom permitted her to make appointments for them to visit. I was the exception. When I asked him about this, he said that none of them really understood what he was going thru. Their visits were either a duty or motivated out of self-interest and not real caring. Some were in denial and offered endless solutions and miracle cures. They all stood around the bed like a bunch of vultures waiting for him to die. Their conversation was shallow, insincere, boring. They were always grim and had no sense of humor. He said I was the only person who would laugh with him and share the medicine pipe and some honest conversation.

Every day when I arrived for our visit, we enjoyed a ritual. I had made a marijuana pipe from deer antler for Pieter to keep next to his bed. I would fill the medicine pipe and fire it up. Pieter and I would share two or three bowls of marijuana, then relax and share the afternoon and evening. It was during these times that I first witnessed how much difference the marijuana makes for those dealing with cancer, chemotherapy and mortality. It buffers the physical discomfort of the disease and the therapies, improves energy, keeps the spirits elevated, stimulates appetite, and promotes understanding.

When I arrived each day, Pieter's girlfriend would meet me as I entered, with a briefing on his present mood and condition. Usually she would say that he hadn't eaten all day, and often that he was depressed and gloomy and uncommunicative. I would ask her to prepare some light food, usually homemade bread with sliced fruit and cheese. Then I'd join Pieter and light the medicine pipe. Within moments Pieter would be laughing and joking and nibbling at the finger foods and teasing his girlfriend. In the worst circumstances, marijuana always comes through and makes you feel better in body, heart and mind.

One of the most important things I've learned while serving critical and terminal cancer patients is that, while you cannot change the fact that you have cancer and must endure harsh treatments, and perhaps meet your death, you can change the quality of the whole experience, reduce the fear and suffering, promote understanding: add marijuana to your protocol. It is more than medicine. It is a friend.

As Pieter's health declined and his body wasted away, we continued to enjoy the closeness of best friends, savoring each day, each moment. The marijuana aided me in understanding and empathizing with Pieter's situation; it helped me stay relaxed and kept my spirits up, so I could be there, with Pieter and for Pieter. It kept our vision clear and helped both of us understand that death is an inevitable part of the life cycle, and this understanding countered much of the fear and apprehension. I heartily recommend marijuana therapy for the loved ones and caregivers of cancer patients. I believe it was instrumental in my being able to maintain a close connection with Pieter during that last year. I'm grateful for that last year and the memories of the daily visits with my best friend. He was a special guy. I miss him every day. One of Pieter's legacies is the knowledge I acquired during that time, knowledge that became part of my thinking as I learned to care for and be a friend to other cancer patients in the years that followed.

Adios, Pieter! Good luck! And happy trails!

WU MING

Age: 75.   Businesswoman, real estate investor, grandmother, retired. Condition: brain tumor.

Wu Ming was my second patient. She is a dear friend. So when she was diagnosed with a brain tumor and scheduled for surgery at University of California San Francisco, I went with her and stayed by her side throughout the ordeal.

During the experience, I met other brain surgery patients and their families. And my sense of caring and wanting to help began to grow.

In the UCSF neurosurgery department, though patients might be counseled individually, they are actually under the care of a team. As I watched the surgery process, I saw 30 patients prepped for brain surgery and operated on by a team of neurosurgeons. All 30 skulls were opened at once, and the team operated on 30 patients at once, under the direction of the chief neurosurgeon. It was an elite and highly trained team on a well-planned mission, with the latest and best of technology and support.

The surgeons moved among the patients. The energy level was very high. The air tingled with electricity. Team members were connected.   There may have been some telepathy going on. There was definitely a lot of adrenaline going on. The buzz was tangible. It was an exciting thing to experience.

After the surgery, Wu Ming was put in the intensive care unit. In the UCSF hospital, the ICU looks like something you might find on the Starship Enterprise.   There are rows of glass cubicles, and rows of registered nurses. Each patient is isolated in a glass cubicle, piled high with technology, screens, lights, wires, tubes. It can be a challenge to find the patient.

A few feet outside the door of each cubicle is a nurse's station: An RN sits at the station by every patient's cubicle, monitoring the information coming from the patient to the rack of screens at the station. If the patient's condition changes by the slightest degree, a signal alerts the nurse who rises and enters the cubicle to make an adjustment. The patient is kept in an absolutely perfect state of equilibrium. It is difficult to imagine more technology or better care.

As I got to know some of the other patients and their families, I learned that one patient, a black woman in her 40s, had no family or friends. From the appearance of her skin, hands, hair, she seemed to be a homeless or street person. Hospital staff said no one had been seen with her, no family, no friends. She was undergoing brain surgery and recovery alone. She was one of those who did not survive. I've wondered many times whether her lonely situation was a factor, and whether I might have done something to make a difference.

And there was a retired man on a fixed income, with a quadrapalegic wife. Being her only caregiver or support, he had hired someone to care for his wife for five days so he could take a bus from Sacramento to San Francisco to undergo brain surgery and return home on the bus to resume care of his wife.

Becoming aware of such desperate and tragic situations made me want to help. It was at that point that I began to look for ways to be a friend to those dealing with serious health issues. That was the beginning of my medical marijuana caregiver practice. At the time, I saw something meaningful I could do. It had little to do with marijuana at first. That became increasingly important, as I learned how helpful it can be.

Wu Ming survived her surgery. A tumor the size of a tangerine was removed from her brain. But she has recovered beautifully. She does not walk in circles or bump into walls or slur her speech any more. Now she hikes long distances and pursues real estate projects and enjoys her grandchildren.

Wu Ming uses marijuana in ordinary ways. She does not smoke. But, for her arthritis and muscle issues, she puts marijuana oil in her cocoa, for an oral dose. And for her joint and muscle pain, she will sometimes use a tincture made with marijuana flower buds soaked in vodka as a topical rub. This is a very old Mexican homeopathic remedy, not commonly used by patients now, but still a well-known way of using marijuana.

I know from experience that it works great on arthritic joints. As it is absorbed into the tissues, it brings a tingling sensation, dulls the pain and improves flexibility. Though, the effects are not lasting, and I must apply it frequently. Unless you grow your own or gather it wild, it is an expensive way to use marijuana.

I'm familiar with two Mexican families whose older members still use this method. They do not purchase or even grow their own marijuana. They gather the *cannabis* in Mexico in places where it grows wild.

Wu Ming also has a special method for using marijuana. Since her brain surgery, she has had difficulty sleeping more than four hours each night. Her doctors want her to have more sleep, to keep other health issues from progressing. Marijuana has a very strong, seductive, intoxicating aroma. One evening, I cut some large flowering stems from the marijuana bushes in my garden and took them to Wu Ming and placed them in her bedroom near the head of her bed before she retired for the night.

I'd selected the strain known as *Medicine Man* (a.k.a. *White Rhino*), because of its refreshing, mint-like aroma and its especially powerful medicinal qualities. That night, Wu Ming found that she slept deeper and longer, that her dreams were pleasantly affected, and that she woke feeling more rested and refreshed than usual. From that time, she likes fresh cut marijuana buds for her bedroom whenever I can manage it.

If you grow your own marijuana, you might wish to try it for sleep therapy. When you cut stems from your plants, hang or spread them to dry in your bedroom, near the head of your bed. Sleep and dream in the "pastures of heaven."

SPRITE

AGE: 21. Student, rock climber, retail health food. Condition: osteosarcoma (bone cancer) survivor; amputee; prosthetic leg; one year of weekly in-patient chemotherapy.

I met Sprite thru her doctor. He called one day to tell me of a beautiful girl who had been living half-wild outdoors in the Sierra mountains with her boyfriend, rock climbing, panning for gold, hiking, skinny-dipping in cold mountain streams and lakes. She was suddenly diagnosed with osteosarcoma, bone cancer, which made it necessary to move to San Francisco to be close to medical treatment.

She didn't know anyone from the Bay Area. Her parents were on the east coast. They were separated. And each had a different kind of substance abuse problem. So they weren't much interested in Sprite's problems. Sprite had no friends, family or support in the Bay Area. No job or income. No shelter. No marijuana. A friend of her doctor had let her and her boyfriend stay in an unused apartment while they were traveling. But it was temporary. Meanwhile, she was scheduled for immediate surgery and a year of chemo at UCSF hospital.

I joined her at the UCSF hospital in San Francisco just before her surgery. Her parents showed up the day after her surgery. Her mother stole her pain medication. That evening, in tears as she lay in her hospital bed with her leg just removed the day before, Sprite told her parents they were making things worse and asked them to go away, to go home and leave her alone.

For the next year, every week for one or two days, she had to check into the hospital for chemo treatments. For convenience, a plastic port is installed in the upper chest of chemo patients, to attach the chemo tube, which drips toxic chemicals into the patient's bloodstream. There is a wide range of toxicity in the treatments. Some chemo patients visit their oncologist's office for an hour to get their treatment, then return home. For patients needing stronger treatments, they keep you in the hospital, often for just a day or two, in case the chemicals are too toxic and cause complications. If the cancer doesn't kill you the chemicals might. The worst case I've seen was kept in the hospital for a full month.

Sprite and her boyfriend stayed in the loaned apartment, recovering between chemo treatments. When she was in the hospital, the boyfriend and I would stay with her. We slept on hospital chairs or made pallets on the floor. I brought marijuana to smoke, and bags of brownies made with marijuana or hashish for eating.

A kind Pakistani donor familiar with my work provided the money for a vaporizer, an electric device that vaporizes and extracts the medicine from marijuana without burning it and creating smoke or odor. The patient gets pure medicine in vapor form, with no smoke. The best one to use is the *Volcano*, at $400. Second best is *Vapor Brothers*, at $125.

In hospitals, there is often oxygen in use. Open flames, as from a *Bic* lighter, could create a fire danger. So patients either go out of their rooms to garden or patio areas to smoke, or they smoke in their rooms with vaporizers or in the bathroom.

The doctors and staff in the UCSF hospital have been open-minded. They have seen the benefits the patients receive from the use of marijuana. And seeing, they have supported its use and have not interfered.

When staying with Sprite for chemo treatments, each time she checked into the hospital, she would put on her gown and sit in the bed hugging her remaining knee and lamenting the loss of the other and moaning how she didn't want to be there and wishing she was anywhere else. But once she was settled in, we brought out the vaporizer and had a few puffs of marijuana.

The next time the nurses came into the room, there might be a Lucinda Williams or Dylan album playing softly on a boom box, and Sprite would be painting the toenails on her remaining foot and talking on the phone, placing an order for sushi or chow mein from a local restaurant, and asking if they deliver to UCSF. If not, the boyfriend or I would have to take a walk.

Nurses would often comment on the mellow, relaxed atmosphere in the room and sometimes wished they could stay. And sometimes they would comment that in some of the other rooms on the floor there is such tension and fear that it would be nice if this atmosphere could be shared.

Only one time have I ever seen a staff member in the UCSF hospital make an issue over the marijuana. Sprite was in her bed and a young nurse came up to her and told her to go in the bathroom to smoke. For some reason she objected to Sprite smoking openly in her room, even though she used a vaporizer.

But Sprite stood her ground. She pointed out to the nurse that her doctor supports her use of marijuana, as part of her medical protocol. She pointed out that she was attached to the chemo rack. And with only one leg she could not very well hop across the room to the bathroom dragging her chemo rack every time she wanted a puff of medication. No thank you.

The nurse was plainly offended at Sprite's attitude and left in a huff. A bit later a different nurse came in to tell us what followed. The offended nurse went to the head doctor and reported the incident in the room. The doctor turned and said to her "Why don't you lighten up?"

Attitudes in the UCSF hospital are the most enlightened and supportive I've encountered. I've never seen any other doctor or staff member at the UCSF hospital object to the use of marijuana or interfere with a patient's use of it.

Kaiser policies are less enlightened, more restrictive. Patients are discouraged from using marijuana in the hospital. And Kaiser doctors generally will not issue a prescription or letter of support for patients using marijuana. Reportedly, some will, in some Kaiser hospitals. But in my experience, I've never met a Kaiser patient who did not need a referral to another doctor to request a marijuana recommendation. At other hospitals in the Bay area, attitudes range from narrow-minded to "look the other way" to tolerant, supportive and even appreciative.

In one Bay Area hospital, the nurses in the children's cancer wing, seeing the benefit one of my own patients received from the marijuana, asked me for a supply of brownies to keep in the freezer for other cancer kids who might be having a bad day.

In another hospital, the head nurse on the children's cancer floor asked me to show her how to grow, bake with, administer and dose medical marijuana. She took it on herself to provide marijuana laced baked goods for the cancer children under her care! God bless her! And all the other staunch medical people who have put the welfare of their patients ahead of their fear of the government.

When she could no longer use the apartment, Sprite stayed with me in the country while recovering from the effects of chemo. Then as she regained her health and strength, she decided to return to the mountains. She rents a house with river frontage, where she swims and pans for gold and floats in the sun in a big truck inner tube. Since her recovery, she has returned to climbing rocks with her new bionic leg, finished three years of college and has a job cooking in a health food restaurant. She is remaining in California State University this fall to continue her B.A. program. She has a life again, and we remain close.

BOTTOM LINE: Patients have a right to use medical marijuana in California, by following the law. Even in the hospital. It is medicine. It is part of your protocol. If your doctor approves your use of marijuana, you can use it. Exercise your rights.

WARNING: Oxygen is always present in hospital environments. Open flames from cigarette lighters present a fire danger. For this reason, hospitals have rules about open flames. But if you are using a properly designed vaporizer, which does not use an open flame or make smoke, there is no reason for objection. Not all vaporizers are the same. Some don't work very well; some break or have other problems. And some could be made that would not be fire safe. I speak only from experience with the two brands named in this book. These have been used safely in hospitals and are efficient choices.

NOTE: Some throat or lung cancer patients, and some patients with chemo sores in the throat and esophagus, have used a vaporizer and report that it feels therapeutic, not harsh.

When patients wish to smoke a pipe or a joint, they will leave their rooms to smoke in a patio or garden area within the hospital. Or use their private bathroom. To clear the odor of marijuana smoke, you can use **Ozium**, a commercial air deodorizer available in small purse size aerosol cans at pharmacies or shops that sell marijuana pipes and supplies. It works. A two second burst will instantly deodorize a smoke filled room or an indoor grow room. It is not a healthy habit or the best of ideas. But in an urgent situation, it can be nice to have it on hand.

If you are in the hospital and wish to use marijuana, go for it. If you are discreet and follow these suggestions, no one should smell or suspect a thing. If staff suspect, they will often look the other way. If you have problems, call Pier 5 Law Offices.

UPDATE: More than a year after returning to her home in the mountains, Sprite was stopped one night by a CHP officer as she was driving home from work. He illegally searched her car and found her medication. She identified herself as a legal patient and showed her documentation. The officer arrested Sprite and jailed her for the night, in direct violation of guidelines in the CHP training manual instructing officers not to cite or arrest legitimate patients or confiscate their medication. The case, though costly, was dismissed.

This is a widespread problem. CHP and other law enforcement throughout the state commonly act outside the law when they ignore the law, abuse patient rights, make unwarranted arrests and confiscations, actions which result in serious loss and injury. In some cases, officers are profiting from confiscations they do not report. In one northern county, some sheriff's deputies are known to have repeatedly confiscated quantities of marijuana that they did not report but sold and profited from.

Some means of sanction is missing. Presently, officers act with impunity. Individual officers and their superiors are not accountable for the loss and suffering they inflict, when they act like thugs instead of public servants and abuse their authority and the law they are paid to enforce. Such behavior by law enforcement engenders fear in legal patients and caregivers and seriously undermines public trust.

STAR

Age: 27; mental age: 10. Downs syndrome. Condition: aggressive terminal leukemia.

Star was the daughter of a refined and cultured family. Having Downs syndrome, she never fully matured mentally. However, she developed other abilities beyond the rest of us. In particular, Star was empathic and to a lesser degree telepathic. She could sense the emotions and sometimes the thoughts of others. If you think about this, you might see that it brings an interesting dynamic to the family circle.

Anytime anyone entered the house with negative or troubled emotions, Star would point it out immediately and would insist that it be resolved or left outside. If a family member, even the pet poodle, was away from the house and lost or frightened or in trouble of any kind, for example with a flat tire or lost keys, Star would sense it and make an issue with whoever might be on hand. She would insist that someone from the house go and find the one in trouble. And she was always right.

With an empath in the family, it became impossible for anyone to harbor ill feelings toward another family member, toward themselves, or even toward the world in general. Living with Star you had to be always emotionally honest and well-balanced. This very fine family group was made finer still by her presence. All the others acknowledge this. And so they were all saddened to learn that Star had an aggressive form of leukemia and would not live more than a few months.

The father of this family is an unusually kind, gentle and caring man. When he learned that his daughter had not long to live, his response was one of the most remarkable and admirable things

I've witnessed. The way this man arranged his daughter's end and passing was as good as it gets. In my long experience in this work, counseling critical and terminal patients and their families, I have not seen any family arrange the passing of a loved one better. This could be a model. Should be a model.

Star's Papa determined to not leave her side. Period. For the rest of her life, if she opened her eyes her Papa was there. And he chose to make the home environment as happy and joyful as could be, not somber. He hired helpers, nurses and housekeepers, who came and went. And when their shift was over and it was time for them to go home, they did not want to leave but to stay and be part of this special experience. Friends came and went by appointment. In that way, Star had a constant trickle of friends and loved ones with her, but never a crowd.

Papa brought a hospital bed home and put it in the spacious family room, near the fireplace. Star's sheets and pillow covers were of sky blue flannel with puffy white Peter Max clouds. Papa brought his own double bed into the family room, so he and Mama could stay close to Star. He brought Star's huge doll collection into the room and arranged hundreds of dolls around her bed. He put an entertainment center, with TV and games and stereo, across the room from her bed.

He sent out for anything Star might want, anytime of the day or night. If she asked for pizza or McDonalds or Baskin Robbins at 3 a.m., he would send someone to get it for her. He could not save her. So he chose to make her remaining time on earth as happy as he possibly could, and her passing gentle.

I had known this family for years. And I had a special fondness for Star. We enjoyed a friendship that is difficult for me to define. She was like a sibling, a younger sister, in some ways.

And she was like a teacher; and an example, a role model. In spite of her differences, she seemed to me a soul more evolved than the rest of us. When she learned how ill she was, she asked her Papa if I could come stay with them for a while. I'm grateful for the invitation. That visit was one of the memorable events in my life. Things I learned at that time have been incorporated into my work with patients since. Thank you, Star.

Her father, knowing of my work, asked if I thought marijuana might be good for Star. Wanting to do everything possible for her, he asked me to try it, so he could observe the effects. A simple request and her doctor provided a letter of support.

I made some marijuana brownies and gave them to her with her desserts after meals. We increased the amounts of brownies gradually. The father was extremely sensitive to his daughter. He watched Star very closely and quickly determined that one third of a brownie with meals put her in a perfect state.

The entire family observed that Star had an improved appetite, slept deeper and longer, enjoyed less troubled dreams, laughed more often, experienced more tender moments with family members.

Once this became clear, the decision was easy. The father committed fully to making the marijuana therapy a part of Star's protocol. This man did everything within his power to make his daughter's last days and passing as happy and easy as could be. Finally, Star passed away in her father's arms. They held one another, and Papa spoke comforting words to her until she was gone.

I admire Star's father. He is an exemplary human being and the best of fathers. Those who know him view him as an unusually kind and caring yet strong figure. In my view, he is one of the most ethical, honorable and noble men I've encountered in life. Being able to witness the way he arranged the passing of his daughter revealed yet another facet of his character. It's a pleasure to know him and to have served his family.

WHITE TARA

Age: 55. Self-employed artist/designer, translator; wife, mom. Condition: terminal breast cancer.

In Buddhist culture, White Tara, a benevolent deity, is a highly evolved form of the female experience. She is born from a tear of compassion shed by a Bodhisattva for the suffering of all sentient beings. She is a healer, a nurse and protector.

I met Tara after she called and said she was referred by one of the doctors who send patients to me for marijuana therapy and counseling. So I made an appointment and went to her home to meet and do some initial paper work. When I got there she confessed that she did not call me because she wanted marijuana therapy. She had heard about my work with cancer patients from discussions in a support group. And she wanted to help.

It seems she was in a group with one patient who had been talking to the others about the benefits she had enjoyed from the use of marijuana. She had described my work and discussed the difference marijuana had made in her cancer experience. So Tara asked her the name of the doctor who had referred her to me. She went to see that doctor and asked for a referral.

I was moved by her sincere desire to help. Tara is a Chinese wife and mother, with terminal breast cancer. She does not want conventional, western medical treatment. She has chosen (as I have) to use only natural and Chinese herbal medicine to ease her discomfort, while she lets nature take its course.

Meanwhile, she wanted to assist my work helping patients. So I asked her to help with the baking. She let me know that she is a world-class cook and would be glad to bake not only brownies but all sorts of wonderful confections.

So I would bring the materials to her house, and together we would enjoy good music and food as we made a crockpot full of marijuana infused butter for her to keep on hand in the freezer, to make brownies, cookies and other things.

There are several women, White Taras, loved ones or widows of patients, who have helped with the baking. This love and caring are felt by all the patients who receive the baked goods. The benefits they've received and the gratitude they've expressed have been endless.

Making Marijuana Butter

Remove seeds and stems, and place 4 ounces of dried marijuana buds, or 8 ounces of leaf trimmings mixed with small buds, in a standard size crockpot. The marijuana should be strained through a wire kitchen strainer or processed through a spice or coffee grinder. Add 5-6 pounds of butter. If you wish to avoid butter in your diet, canola or peanut oil may be used instead. Cook on low setting for 48 hours, stirring occasionally. If the buds are very sticky, cook even longer. If the oil bubbles and boils, turn off heat for a while. When cool, turn heat on again.

When done, strain through 4 layers of cheesecloth. Gather corners and twist tight to remove all the good oil or butter from the marijuana pulp. Discard pulp. When cool, you may store butter in freezer in gallon Ziploc bags. Oil may be kept in the refrigerator in plastic jugs.

Now all the medicine from the marijuana is in the butter or oil. You can use this to bake anything: brownies, cookies, cakes, confections of any kind. Some patients put it on their oatmeal or pancakes. Some like it in sweet potatoes or pasta or salads. It is great in smoothies or cocoa.

DOSE: The average adult dose is 1-1½ tsp. If your body weight is low, from illness or treatments, use less and experiment to find your comfortable dose. Repeat every 4-6 hours.

Here are four of the most common recipes we use:

Gourmet Brownies

12 ounces unsweetened baker's chocolate
1 ½ c. marijuana butter or oil
5 c. sugar (or 4 c. honey)
8 eggs
3 Tbsp. vanilla extract
¼ tsp. salt
4 c. flour (8 c. flour if you use honey)
½ c. Southern comfort (or other bourbon or your favorite liquor; or 1-2 Tbsp. coffee crystals dissolved in 1 Tbsp. hot water)

NOTE: you can omit the bourbon if you don't want the liquor flavor. However, the bourbon is the touch that makes these brownies "gourmet." Don't be concerned about the alcohol. It will evaporate at 170 degrees, during cooking, and the finished product will be alcohol free, leaving only the flavor. Just like vanilla extract.

Heat and mix chocolate, butter, honey and liquor in a double boiler (a steel mixing bowl sitting on a pot of hot water works fine).

In separate mixing bowl, beat eggs and add vanilla and salt. Temper eggs with hot mixture. Then add eggs to the chocolate mixture and stir well.

Add flour and mix again.

Pour into 2 9x12 oiled pans. Bake 40-45 minutes at 375 degrees. Cool and cut into 2 inch squares.

For storage, bag brownies in gallon Ziploc bags and freeze. They will keep indefinitely.

Oat Cakes

Blend:

2 c. marijuana butter or oil
2 c. honey
2 c. applesauce
1 c. brandy, bourbon or other booze (or rice milk)

add:

3 c. flour
6 $\frac{1}{2}$ c. rolled oats
2 c. raisens or dried cranberries
2 Tbsp. vanilla extract
4 tsp. baking powder
1 tsp. cinnamon
$\frac{1}{2}$ tsp. ginger
$\frac{1}{4}$ tsp. salt

Mix and pour into two 9 x 12 inch oiled baking pans. Bake 40-45 minutes at 375 degrees. When cool, cut into 2 inch squares. Bag and store same as brownies.

NOTE: the addition of liquor (optional) in this recipe makes the finished product reminiscent of fruitcake.

DOSE: Dose for brownies or oatcakes is 50 grams, or one 2" square, for a 170 lb. adult. Experiment to find your best dose.

Hot Chocolate

In small saucepan place 1-1½ tsp. marijuana butter or oil. Add 1 rounded Tbsp. Scharffenberger's gourmet cocoa powder (or 2 wedges Cibarra Mexican sweet chocolate). Mix over medium heat until you have a thick chocolate paste. Add rice milk, soy milk or cow milk. Heat to taste. Sweeten and enjoy!

HASHISH: Instead of marijuana oil, you may use plain cooking oil, adding 1/4-1/3 gram of hashish. Heat oil and hash, stirring until hashish is dissolved. Then proceed with recipe.

This medicinal cocoa beverage usually accompanies my breakfast. And I like to enjoy a cup of marijuana cocoa in the evening, before bed.

Fruit Smoothie

In blender place 10-12 ounces fresh or frozen berries or other fruit. Add 1-1½ tsp. marijuana oil, 1 Tbsp. honey or other sweetener, 3 scoops soy powder. Cover with soy milk. Blend until smooth. Enjoy!

Tara, you are one of the lights in my life. You have helped with the baking and have been a blessing. You have lightened my load and freed me to spend more time with patients. The patients who received your baking send their love, their thanks and their prayers.

JACK & JILL

Age: 62. Self-employed carpenter, handyman, painter, cabinetmaker. Condition: Hodgkins survivor.

Jack and Jill are a close couple. When one of a couple is diagnosed with cancer, both are thrown into a crisis. The spouse, though they may not have cancer themselves, is always overloaded and overwhelmed. There is not only the normal work of one's job and keeping a household, but now there's a sick person to care for, and phone calls and paperwork and medical appointments and insurance and ordering the right medications and administering the right medications and on and on. Being the loved one of a cancer patient is exhausting. And the worry and stress and loss of sleep take a toll.

In Jack and Jill's case, they had the added burden of not being able to live in their own house during the cancer crisis. They live in a redwood forest north of San Francisco, in a hand-made house that Jack built. It was an unusually hard winter and the dirt and gravel road to their home was washed out. So the couple had to find other accommodations while dealing with Jack's cancer.

They stayed with friends all over Marin and San Francisco. And the Red Cross gave them money for hotel rooms when friends' homes were not available. Dealing with cancer is hard enough. But it's a different picture when you're homeless. At least Jack and Jill have friends to help. Most homeless folk don't.

On top of all else, with Jack unable to work, their income was reduced. Jill does bodywork, so she continued to serve her clients, in addition to everything else on her plate. They could not afford the loss of income.

Initially, Jack was treated at Marin General Hospital and California Cancer Care in San Francisco. His cancer was cured. And they were very happy, especially Jill, who had been living long with the fear that she was losing her husband and best friend. She was ecstatic to learn that he was cured and cancer free! That is relief that only cancer patients will understand.

It was a heavy blow, therefore, to learn from follow up tests that the cancer had returned, aggressively. They were terrified. It is wonderful to survive cancer. But it's a terrible thing to learn that it has returned.

One patient, a martial arts master, has been through the cancer surgery and chemo experience four times. Though he is among the strongest, most courageous people I know and a real warrior, even he dreaded the follow up tests and consultations. Even he was discouraged to learn that the cancer had returned. It is crushing to learn that the harsh treatments you've endured have not done the job and you must begin again.

I'm happy to say the four-time cancer patient is a survivor. He's tested cancer free for several years. He's returned to teaching martial arts five days a week and traveling on weekends to officiate at tournaments. He's a very busy person. Though, I know he constantly wonders whether the cancer will return.

He parties a lot, with family and close friends. He organizes picnics and barbeques and banquets; he goes to concerts and dances. He celebrates in every way imaginable, as often as he can. And he travels with family groups, as often as he can. He lives as though the cancer might return at any moment. And he wants to squeeze everything he can from every moment he's alive. Cancer changes forever one's goals and values. Survivors often wish to do something to help others. And they often find a less rushed, more leisurely routine, and learn to savor life instead of rushing through it. Enjoy this day. You may not have another. It's a commendable approach for all of us:

Take the cup and the jug
Sit at ease in the green field at the edge of the stream
For the wheel of time soon enough
Will bear you to the sea of extinction!

Omar Khayyam
"The Ruba'iyat"

When Jack's cancer returned, the prognosis was grim. Several times I found Jill quietly weeping where Jack would not see. There were not many options left. Jack's oncologist finally chose to try a bone marrow transplant with an experimental and highly toxic chemo treatment. Jack was admitted to the UCSF hospital for a month.

The chemo treatment used on Jack was so toxic that they put a sign on the door to his room warning anyone entering the room not to touch the patient or their own eyes or mouth and to wash their hands before leaving the room. That's some scary stuff to be pumping through your body.

As the chemicals passing through Jack's body sweated to the surface of the skin they would eat the skin away unless the patient took 2-3 baths every day. As I watched Jack endure this treatment his agony was plain to see. Sitting in his bed, he constantly moaned and hugged himself, rubbed his body and rocked in his bed, incessantly. He could find no ease. Exhausted, he could find no sleep.

Another patient, a kind-hearted person with an antique business on Union Square, donated money to buy a vaporizer for Jack. Jack smoked marijuana both before and after the bone marrow transplant. However, the bone marrow transplant patient is extremely vulnerable to infections, especially of the respiratory system. For this reason, Jack's oncologist restricted use of smoked marijuana for the month of the transplant.

Jack has frequently expressed to me how much relief he gets from the marijuana. Without it he can't eat. Nausea is ever-present. If he forces food down, it comes back up. He can't sleep. Fatigue mounts. Physical body pain is constant and unbearable. But with the marijuana, he has an appetite. He eats and keeps his food down. He sleeps. And his body pain is reduced substantially.

On learning that he was to spend a month in the hospital undergoing a very hostile chemo routine, without being able to smoke marijuana, he became panicky. He begged me to not forget him. He had been on both a smoking and an eating therapy. I promised to bring marijuana-medicated edibles to the hospital for him, so he could continue his oral therapy during the transplant and chemo.

Jack had tried *Marinol*. But as with every chemo patient I've known, he found that it was not as effective as the real thing. Not a single chemo patient in my experience will use *Marinol* if they can get natural marijuana. Some will not use *Marinol* even if they **cannot** get the real thing. I've used *Marinol* and find it to be only fractionally effective and the effects crude, heavy and sluggish by comparison to marijuana. And it costs $17 a dose, much more than natural marijuana.

Jack needed strong hash cookies. So Tara and I made a special batch of cookies for Jack. Instead of marijuana oil in the recipe, we used plain canola oil and dissolved hashish in it, 1/2 gram per cookie. This strengthens each cookie by a factor of 4 or 5. Normally, when I want to increase the strength of the baking for chemo patients, I will add 1/4 to 1/3 gram of hashish per dose. But for Jack, I wanted these cookies to be super strong. These little tiny hash cookies, maybe an inch and a quarter across, did the job nicely.

CAUTION: Doses of hashish this high are for chemo patients only. They buffer extreme levels of physical discomfort and the side effects of chemotherapy. They help the chemo patient get away from the body or achieve sleep. Even then, do not eat a whole dose the first time, especially if your weight or strength are low. Experiment by nibbling a little piece every 60-90 minutes, until you find your comfortable dose. If an experienced recreational user were to eat a cookie with 1/3 to 1/2 gram of hashish, they would very likely find themselves too heavy to stand or sit and possibly falling into deep slumber. Do not worry. It will wear off. Marijuana is not lethal.

NOTE: The estimated L-50 dose for marijuana (L-50 is the dose of a given substance that would be lethal for 50% of adults) is 40 pounds of marijuana consumed in a single dose, a feat which is obviously impossible. **Marijuana is not lethal.**

Marijuana and hashish confections work exactly like *lembas*, the waybread of the elves: "They will serve you when all else fails." They soothe your pain, bring courage to your heart, lift your spirit, and give you energy that "will keep a traveler on his feet for a long day of labor."

The hash cookies were perfect for Jack's extreme level of discomfort. They can buffer the side effects for even the worst chemotherapy! I've seen it again and again.

If you ever have to undergo chemotherapy, I urge you to add marijuana or hashish to your protocol. **You cannot change the fact that you have cancer or that you must endure a harsh** therapy.   But you can change the quality of the experience, mitigate the discomfort and reduce the fear and suffering. You can buy some hashish at a medical marijuana co-op and ask a friend or loved one to make you some hash cookies, following the instructions in this book.

With the oral dose of hash cookies, Jack was able to get through the transplant and chemo. He is now finished with treatments and is testing cancer free. He and Jill are still living with friends in Marin. But I spoke with them on the phone earlier today. They just received word that the road to their home is repaired!   After months of living like itinerant vagabonds while dealing with cancer, they can finally return home tomorrow! Cancer free! Good luck, you guys!

SKEETER

Age: 28. Logger, gold miner, professional caregiver, marijuana farmer, husband, father. Condition: busted.

When I met Skeeter, he was the boyfriend of one of my patients, Sprite. They were 21 then. He had known her only three months. They were living outdoors in a tent in the Sierra, climbing rocks, panning for gold, skinny-dipping in cold mountain rivers and lakes: two kids enjoying summer in the mountains. Then she was diagnosed with cancer and scheduled for immediate surgery and a year of chemotherapy at UCSF hospital in San Francisco.

Many guys Skeeter's age would have just said "So long! It's been good to know ya!" and carried on with their summer in the mountains. But not Skeeter. He stayed with his friend thru her crisis. She had no family or friends in the Bay Area. Had no resources or support of any kind. She was essentially homeless, an east coast kid enjoying a summer in California. Suddenly she was facing a year of cancer treatment and therapy. Alone. I never saw Skeeter waver. He never considered abandoning her. He's a staunch buddy in a crisis.

For a year, Skeeter lived with her in the hospital and a variety of temporary accommodations. In the hospital, he slept on chairs or a pallet on the hard floor. He fixed her food, shopped and ran errands for her, helped her bathe and dress, walked long distances to health food stores to find special foods for her.

Over the years that I've been doing this work, I've seen many people deal with the cancer crisis. Some are heroic, others less so. Tragically, I've even seen an affluent, educated Marin county family cast their beautiful 21 year-old daughter out on the street to suffer alone because she had hepatitis C and they were afraid of contagion.

As I watched Skeeter respond to the cancer crisis, I developed enormous respect for him, for his integrity, loyalty, courage. So when he came to say that he wanted to be a caregiver and asked me to show him how to grow marijuana and cook with it and administer to sick folk, it was an easy decision. Besides, with my age and poor health, a pair of strong, young arms around the farm would be welcome.

Skeeter and his girlfriend came to live with me for a time, during her year of chemo and recovery. Skeeter cut the firewood and fixed the fences and helped in the garden. And he learned how to grow marijuana. In fact, now he grows bigger and better marijuana bushes than his teacher.

Soon as they could, the young couple moved back to the mountains. Skeeter became his girlfriend's caregiver and began to grow marijuana for her. Others in his community learned of his activity and asked him to be their caregiver. The local co-op, which provides medical marijuana for the entire community, asked him to share his medicine with them. So he has donated a share of his crop each year to the co-op. Some is sold to support the co-op; some is given at no cost to patients who cannot afford it.

Over time, Skeeter and Sprite grew in different directions and each found a new mate. Skeeter married and had a baby. He and his wife bought a few acres with an old frame house and settled down to make a life. Their property is forested with evergreen trees: cedar, pine and fir. Skeeter logs and sells some of the trees. Others he mills into lumber, which he sells to local builders for income to support his young family. Several times each year, he takes backcountry trips to search for gold.

Mostly he does this for solitude and pleasure. But he finds enough gold to make the trips profitable, as well.

In addition to prospecting and logging, Skeeter grows a big vegetable garden for his family and neighbors, keeps chickens for eggs, beehives for honey, and he grows a big marijuana garden for patients, the co-op and the clinic. Another thing he has learned from me is the joy of sharing produce from your garden with others.

I find Skeeter to be a hard-working young man, unusually kind, caring, compassionate. And he's an extremely honest and ethical person. He stands tall, has a giant handshake and a clear, steady gaze. I'm proud of him and regard him as among the finest young people I know.

So it is with great sadness that I've just learned the local sheriff and the feds have arrested Skeeter and his young wife, taken their baby into custody, ransacked their home, destroyed their medical garden. Their families have had to spend money on bail and lawyers. Their lives and those of their extended families have been permanently affected. The authorities prevent their effort to have their baby returned. Instead of showing compassion, the woman in charge of Child Protective Services in their county expressed meanness: "You won't have your baby back for a long time. And by the time this is over everyone in both your families will hate each other." Their home and vehicles have been seized. They and the patients they serve have no medicine for the coming year. Skeeter will probably have to spend some time in federal prison. How has this made the community safer? This kind of thing happens every day. Your tax dollars paid for this and you have a share of the karma.

Though medical marijuana is legal in California, attitudes are different from county to county.   In conservative areas, particularly in the central valley and the Sierra, the eastern district of California, and in northern and southern counties, as well, a provincial mindset is common and local law enforcement and prosecutors still cooperate with the feds to harm patients, though their salaries are paid by Californians.

The most law abiding and therefore safest areas are San Francisco and the coastal counties, from Santa Cruz in the south to Humboldt in the north.

In the town of Fairfax, in Marin County, local law enforcement protects medical marijuana patients. I have seen them insist that CHP or police from neighboring towns leave Fairfax and quit harassing patients they followed and stopped in Fairfax. God bless them.

In Mendocino County, attitudes are very relaxed and marijuana has been welcomed. Reportedly, within the city limits of Ukiah, so many citizens were growing marijuana that the distinctive aroma was becoming inescapable anywhere in town.   And some citizens complained to the district attorney's office.   When questioned about the issue by a local reporter, the D.A. is quoted in the local paper as saying "Some people have too much time on their hands."   But an ordinance was established: marijuana grown within the city limits of Ukiah must be grown indoors.

HOMELESS SOULS

Sleeping somewhere
Different every night
My dream is always of Home
~Anon.~

Late one night, as I left a hotel on Van Ness Avenue in San Francisco, where I had just attended the annual NORML conference on marijuana, I was walking down a narrow alley and stepped off the sidewalk to avoid a pile of boxes and debris next to a dumpster. As I passed the pile, I heard a noise and realized that this pile of debris was someone's home. There was a person in there hiding from the cold night and all the devils that roam the streets at such an hour.

I had a small bag of marijuana with me that I'd just been smoking from in the hotel at the NORMAL conference. I took it from my pocket and put a lighter and packet of rolling papers in it. I opened a flap of the cardboard and tossed the bag inside. The medicine would ease that person's physical discomfort, cut the chill, ease their heart and mind. They'll feel good for a while and then sleep better than they have in a long time. When they wake, residual effects will make tomorrow brighter. I wished I could do more. Then I realized I could!

This was the birth of a new mission: Every now and then, I began to take pleasure in filling my coat pockets with baggies, each containing 1/8 ounce of marijuana, a small pipe or book of papers and a lighter. I would walk through the city night, in neighborhoods where homeless folk are often seen living in the sheltered doorways of shops or in cardboard boxes in alleys and side streets. As I passed, I would drop one of the baggies and keep on going. Sometimes I give hot coffee or Chinese soup on a cold night. But a gift of marijuana takes caring a level farther. I know I'd rather have a bud than a cup of coffee. I might give away an ounce or two in a night, in amounts of an eighth ounce. It is a small thing to do. But there are a few homeless folk who will have a better night thanks to a little gift of marijuana.

I realize this is a questionable practice. However, I find that a majority of the homeless souls I encounter have cancer, epilepsy or some other serious or chronic health issue. And for those who may not have a diagnosed health problem, a little marijuana certainly lifts their spirits. Marijuana is a benign and gentle substance. I've seen the benefits these little gifts bring to the homeless. And I've discussed the activity with two doctors. Neither could find reasons to discourage me.

For some homeless patients, I've maintained close relationships over years of time. In addition to the gifts of marijuana, I bring food, clothing and books when I can. I am the only person who visits some of them or remembers their name. I'm the only person who brings them any kind of gift. The system has forgotten many of the homeless on our streets. As a priest it seems an inescapable duty to do what I can to ease the loneliness and suffering of the homeless souls I encounter. And gifts of marijuana have made a difference.

There are sick and suffering folk all around us. There is a huge need for medical marijuana caregivers. Would you like to make a difference? Does your karma need help? Be a medical marijuana caregiver. Are you with a church or other group? Organize a collective. Establish a garden. Grow marijuana and give it to sick people. You will find your first patients in your own group, or among your own family or friends, among the homeless in your own community, in the cancer wings of your local hospitals.

SMOKEY

Age: 68; Deputy Sheriff; National Park Ranger. Retired, disabled. Condition: advanced diabetes, with complications; crippled, amputee, wheelchair bound.

I learned of Smokey from his doctor. When I encounter a patient who has a doctor who will not write a prescription for medical marijuana—Kaiser patients are a good example—I refer them to a list of physicians who are willing to write prescriptions in support of medical marijuana patients. Smokey's doctor is one of these.

The federal government has physicians unnecessarily intimidated, afraid to act and think freely with regard to marijuana, despite complete legal protection for doctor/patient consultations.

Doctor John is on my referral list. He is not an activist. He is a quiet, conventional, old-fashioned physician and grandfather. He thinks inside the box and is not the turkey to raise his head above the flock.

Years ago Dr. John had occasion to use marijuana on himself to control a painful condition that no other medication would affect. He had an attack that was accompanied by a severe, debilitating headache and nausea that produced vomiting. Nothing in his black bag had any effect.

Knowing that marijuana has an effect on nausea, he smoked some marijuana, and immediately his nausea and vomiting ceased and his headache began to soften. He was impressed not only with the effectiveness of the marijuana, but also with the speed of its effect. He began prescribing it for some of his patients.

One day, he said, "One of my patients is very poor. He is a man with no family, few friends, no income, no resources of any kind beyond public assistance. He is crippled and is wheelchair bound. He lives in an urban convalescent home. He uses medical marijuana when he can get it. I believe he spends his public assistance income for it. It keeps his spirits up, in spite of his depressing situation. He could use a friend."

Doctor John wrote a marijuana prescription with the patient's name and the address of the hospital where he lived. The next time I was in the neighborhood, I stopped by to meet him.

It was late morning on a sunny day. After I parked in front of the hospital and was walking toward the entrance, I spied a thin, bald-headed old man sitting in a wheelchair in the sunshine at the end of the parking lot near the dumpster. He chose to be outdoors (that's what I'd do), instead of being cooped up inside the hospital. He has a good attitude and seems to be playing a bad hand well.

When I came within voice range, I called "I'll bet your name is Smokey!" Coming from a stranger, that got his attention! "You'd win!," he called back.

As I drew close I said "I'm Chemo Sabe, a Buddhist monk. I grow marijuana and give it away free to sick folk. Your doctor told me about you and said you could use a friend." I held out my hand and he gave me a huge handshake for a skinny old man in a wheelchair.

I asked him if he'd like me to be his caregiver. He said "Yes, sir! And god bless you for it!"

"Do you have any marijuana right now?" I asked. "No, sir," he replied. I pulled a small jar of *Blueberry* buds from my bag and handed it to him. He had a small corncob pipe and a lighter in his pocket. He immediately filled the pipe and offered it to me. But I make it a habit not to smoke or drink after patients. So I declined the pipe, produced a joint from my pocket and fired it up while Smokey lit his pipe. We moved from the sun to the shade of the hospital building, Smokey in his wheelchair, me seated on a stone planter near the hospital entrance, puffing the medicine plant and watching people.

Where did you get the name "Smokey," I asked? "It's because of my background in law enforcement. I was ten years a deputy sheriff and ten years a national park ranger. But I'd like to think it's because I sit in my wheelchair smokin' 24/7," he grinned.

Smokey can be found sitting in his wheelchair in the parking lot in front of the hospital 24/7, smoking his corncob if he has any marijuana, and watching the world go by. He likes to be outdoors. At night it is generally cold, and the skunks always come out to see what might have fallen from the dumpster. So Smokey wheels himself inside the entrance to the parking garage, just far enough so it isn't too cold and the skunks have their space, but so he can still see and feel the outdoors.

As I got to know Smokey, I learned that he receives $799 a month from Social Security. Of that he pays $776 for his room and board at the convalescent hospital. The remaining $24 he spends for marijuana at the local co-op. It buys a tiny amount, not enough medicine to last half a week, much less the month. And it leaves nothing for other essentials, like clothing.

When I'm in his neighborhood, I'll stop and sit with Smokey a while and watch the world go by. I might bring deli sandwiches or Chinese noodles to eat. Sometimes I bring music CDs for him to play on his boom box. Or a warm sweater or shirt without holes, left behind by a patient who has passed on.

We talk about all the mundane things that people talk about. But we most like to talk about marijuana. About how much 40 years of breeding has improved the strength and quality of marijuana. About the different strains and flavors and aromas and effects of marijuana.

Smokey may have lived his life as a deputy sheriff and a national park ranger. But he knows his marijuana. He likes to tell you that cops always smoke the best pot. From the bags they confiscate, they choose the best for themselves to take home.

A patient who is a prostitute says the same: "Date a cop. They always have the best pot and the best pills!" Ironically, the cops she dates are providing some of her medical marijuana!

Some strains are named for their aroma or the flavor they leave on the tongue: *Blueberry, Juicy Fruit, Bubblegum, Mango, California Orange, Grapefruit, White Chocolate, Honeydew, Cherry Bomb, Strawberry Web.* These sweet, refreshing flavors and aromas are pleasing and begin to make you feel good as soon as you crush the bud, even before smoking.

When marijuana smokers are together and the conversation goes to marijuana, it can sound like wine buffs talking about their latest discoveries. Just like the wine fairs in California, there is an annual marijuana fair in Amsterdam. The best breeders and seed companies from all over the world bring their latest achievements to Amsterdam to be tasted and judged in the *Cannabis Cup* competition. And similar to wine, they are judged for flavor and aroma, as well as effects. Awards are given for the best strains, the best indoor, best outdoor, best sativa, best indica, best hybrids.

Nowadays, experienced marijuana smokers can break open and crush a marijuana bud for visual and olfactory inspection; take a puff or two for flavor and effect; and then tell you whether it was grown indoors or outdoors, fed chemical or organic nutrients, whether it is sativa or indica or a hybrid.   And connoisseurs will often tell you the exact strain and, sometimes, what part of the state it was grown in.

One evening as Smokey and I sat puffing and musing, a frail, skinny, stoop-shouldered, squint-eyed granny, wearing slippers and shawl, approached from the direction of the hospital using a walker to slowly make her way.   As she drew near, I gave Smokey a heads up and cautioned him to put his pipe away.   He turned to look, then responded by saying she would probably appreciate a puff.   Sure enough, when she reached us Smokey offered her his pipe.   Her eyes slowly got big and round and a wide smile stretched across her toothless face.   She sat down in her walker, took the pipe and drew in a hefty lungful. "Blueberry?" she said. "Yes'm," I replied. Precious moments.

Other patients at the hospital where Smokey stays have become aware that he has a caregiver who brings free marijuana.   Individuals have asked in different ways whether I might accept another patient.   This is common.   When those close to one patient see the benefits available from marijuana therapy, they want those benefits, too.   And they have a right.

It would be great if the 200 poverty cases in this convalescent hospital had enough marijuana to smoke.   Yet, I'm just one caregiver.   Sometimes, I feel like Ma Joad felt in Steinbeck's *Grapes of Wrath,* in the migrant tent camp when a group of hungry children stood hopefully watching her cook a pot of soup for her family.   Ma felt helpless because she could not feed all that were hungry.

THE BROTHERS GREEN

Ages: 80 & 85. Farmers, retired. Condition: gall bladder, arthritis, insomnia, gout, aches & pains, old age & sour grapes

Bud and Herb Green have been farmers all their lives. They are brothers. Though they are five years apart in age, they had been living together on the family farm in Michigan so long that they have become like twins. They even dress alike, in cotton work shirts, denim overalls and straw cowboy hats. They walk and speak alike and even finish sentences for one another, like an old married couple.

They had not seen much of the world since the depression. Their parents died, leaving the family farm to them. The family had been there for several generations, farming the same piece of ground, raising vegetables and chickens and eggs and honeybees and all the other things they needed for themselves. They cut their own firewood for winter heat. And they grew a small crop of orchard fruit, which they sold for expense money. They baked their own bread and made their own wine.

They were self-sufficient, able to produce the essentials they needed to survive, seldom having to leave their own property. Their bank was an old sock hidden in the bottom of their tackle box. They didn't have TV. They didn't take the newspaper. Bud and Herb lived a quiet peaceful life, for eighty-plus years, without much contact with the modern world. Time stood still for them.

Then one day an engineer advised the state that the farm belonging to the brothers Green would make a good route for the new interstate highway. Bud and Herb had never seen an interstate highway. They were still driving their antique Dodge truck on dirt roads, and shooting deer over the hood when they got the chance. When a guy with a bulldozer showed up and started knocking down trees in their orchard, Bud gave him a warning and then peppered him with rock salt from his old shotgun. He had planted those trees bare root and tended them all their lives. Who was this madman on the bulldozer?

The bulldozer driver found himself back in time about 100 years, the rock salt beginning to sting. Shotgun shells loaded with rock salt instead of lead pellets is an old country remedy for thieves and other pests that you don't necessarily want to kill. Though, they'll remember which farm to avoid. Not equipped to deal with an armed attacker or rock salt, the bulldozer driver ran off, leaving his bulldozer behind.

Pretty soon, the riot squad came crawling up, surrounded the farm and contained the situation. They arrested the old man for assault with a deadly weapon and ran him through the system, judged him mentally incompetent and locked him up in a mental institution. Welcome to the future, Bud.

Bud's daughter is a friend of mine, living in San Francisco. She went to Michigan to petition the court to release her father into her custody, with the assurance that she would sell the farm and take him to live with her in California.

The request was granted. The daughter sold the farm and brought Bud and Herb to live in California. With the money from the sale of the farm, and some more donated by friends, she was able to buy at auction a small country cottage on a single acre, with some river-frontage. Bud and Herb were happy here. They grew a large vegetable garden and fished in the river. They had their social security now and did not need to raise an orchard crop any longer. Life was pretty good. Then Bud's gall bladder went bad.

Bud had been a wine drinker. There was always a gallon jug of "100% Pure Wine" sitting on the round oak table. He offered the jug and a clean jelly jar to every visitor. He sipped wine all day and a little more heavily in the evening, sitting by the cast iron wood stove. Yes, life was pretty good.

But then the gall bladder gave out and had to be removed. Which meant that Bud could no longer drink his wine. And that was a serious matter. Life was less good. Without the wine, Bud was unhappy, cranky, didn't sleep well, the arthritis was troubling him, painful gout flared up, digestion was bad.

So one day I gave Bud some marijuana seeds and advised him that with these he could grow some medicinal plants in his garden that would replace or exceed the benefits he had gotten from the wine. So he planted the marijuana in his garden, behind a patch of corn, which shielded the marijuana from view from the front of the property.

Being an experienced farmer, Bud raised a fine crop of marijuana. His plants were 12 feet tall, and the large sun leaves on the lower parts measured 15 inches across! Once the first crop came in and Bud began using it, he wondered why he'd ever been a wine drinker. Instead of his gallon wine jug on the table, he now keeps a large Half & Half tobacco can full of marijuana. Now, instead of wine, he offers the marijuana to guests. He fills his old briar pipe with marijuana instead of tobacco. And life is good again. Or was. Until the fight.

One afternoon, Bud and Herb got into a fight. No one remembers why. But these two old guys in their eighties were out in the driveway in front of their house yelling and throwing empty gallon wine jugs at each other, that Bud had saved and stored under the front porch. The jugs were breaking in the driveway and street. The noise of the yelling and the breaking glass caused neighbors to call police.

When the sheriff arrived, a deputy spied marijuana sticking up above the corn! The corn was 12 feet high! But the marijuana was higher! Bud and Herb were arrested and charged with cultivation. Their marijuana was cut down and confiscated. It filled the rear seats of four sheriff's cars.

At the arraignment, the judge asked Bud why an eighty-year old man was cultivating marijuana. Bud explained about the wine and the gall bladder, and how the marijuana was a good substitute.

The judge did not think these two old guys were drug lords. If they weren't medical, they should be. That much was evident. So he dismissed the charges, with the warning to Bud not to grow any more marijuana. Bud responded by saying "I'm an old man. I don't have much time left. Life is not so good without my wine. The marijuana makes things better. Your deputies have destroyed my garden. I'm going home now to plant some more."

MARY JANE BUSH

Age: 52. Corporate lawyer. Condition: ovarian cancer survivor.

Mary Jane is a cancer survivor. She has been through the cancer experience and returned to her life as a corporate lawyer and head of a household. With one notable improvement: She is now a marijuana farmer!

Mary Jane, like so many cancer survivors, continues to smoke marijuana to keep herself in good health, as a preventive, to keep the cancer from returning. She finds that a few puffs on the medicine pipe after work helps her relax and quickly shrug off the tensions of the day. She enjoys preparing dinner, and finds that it tastes better and her appetite is improved. She enjoys the evenings with family and friends more thoroughly. She sleeps better and wakes up feeling refreshed.

Mary Jane has found such relief and benefit from marijuana that she decided to learn how to grow it for herself. She asked me to show her how it's done. Over the years, I've shown several patients how to grow their own marijuana.

For years now, Mary Jane has grown three plants each year. At first, near the end of her treatment, when she was still weak and unable to do any physical work, I instructed Mary Jane, while she directed the labor of her friends and family.

They dug three big holes, four feet in diameter and three feet deep, eight feet apart. These were filled with bags of *Foxfarm Ocean Forest Organic Potting Soil*. If your local nursery does not have it, a search will be rewarded at harvest time. In the center of each prepared hole, they planted a single marijuana plant. The strains were always award-winning strains from

Amsterdam, obtained at a neighborhood co-op. Some co-ops provide live plants and/or seeds, as well as medical marijuana. If you cannot find live plants, you can grow from seed.

You can save any seed you find in the best buds you consume. Some co-ops sell seed. Buy your seed legally at co-ops, if possible. You can often select the genetics and strains you want. Choose award-winning strains. Prices for ten seeds are $100 in the co-op; can be as high as $300-400 from a catalog.

This may seem high, but it's about the same as the cost of live plants. It can cost much less if you use a seed to grow a mother plant and then make dozens of your own cuttings from the mother. This takes longer and is more work than buying started plants. But it is less costly. And it is enjoyable and satisfying.

Although seed can be ordered by mail from online seed catalogs outside the USA, it is a crime. The feds watch the mail. It is dangerous. If you choose to purchase seed by mail, don't use an address to mail seed where you will have your garden. It could start an investigation. And it is better to be safe than sorry. Buy seed legally, if you can.

Generally, I select *Blueberry and Medicine Man* (a.k.a. *White Rhino)* as the hybrid strains for the big plants grown in my garden. These are fine for northern latitudes. They will do fine in the south, as well. And these two particular strains seem to have better results with a wider range of health issues than other strains I've tried. They have greater than average THC content, with strong narcotic effect, for the physical needs of chemo patients; cerebral effects are soaring, not lethargic; and the effects last long, six hours. I've always come back to them and have finally settled on these as having superior medical value. They are both award winning strains and have been around many years. My long experience with them, together with the fact that patients repeatedly ask for these two strains over others, may influence my choice. They are both easy to grow, in both northern and southern latitudes, and will produce a good crop for you. Though, they are not good choices for indoor growers.

Do not select Sativa dominant strains for northern latitudes. It will not finish. Select strains that are 60-80% Indica, i.e., *Blueberry, Romulan, Medicine Man, Snow Cap,* etc.

In addition to the dozen large hybrids, I also like to have two or three dozen 100% Indica plants--such as *Mangolian Indica, Purple Kush, Northern Lights* or *Super Skunk*--in 7-10 gallon nursery pots. Indica has a stronger narcotic effect than Sativa or hybrids. And it is, therefore, more desirable for chemo or radiation patients with the greatest discomfort.

Indica is also the best choice for indoor growers. It has a short, fast cycle and produces compact plants. If you veg Indica only 4-5 days, then go to flowering light (12/12), your plants will finish in 7-8 weeks, at 12-16 inches tall, and will yield 1 ounce each (dried) of sticky, skunky Indica. Each plant will require 1 square foot of space on your table.

NOTE: There are probably no more than 100 people in the world who possess 100% pure Indica strains. But for convenience, I use the term as it is used in magazines, reference materials and seed catalogs, to refer to the predominately Indica strains that we are familiar with.

Indica plants will fully mature earlier than the larger hybrids, like *Blueberry* or *Medicine Man.* Indica tends to be a small compact plant, grown in the ground, and it will ripen and finish just fine in a 7-10 gallon container. Sativa and the hybrids need much larger containers or to be in the ground.

Food, water and light being equal, most strains of marijuana will ripen and finish 2-3 weeks earlier in a container than if planted in the ground in the same location.

Therefore, growing a group of Indica plants in containers gives an early crop. First plants to ripen are the Indica plants in the containers; then, the large hybrids grown in the ground. This gives two separate crops about 3-4 weeks apart and conveniently spreads out the work of harvest and trimming and drying. As a rule, plants grown in containers will yield approximately one ounce of dried marijuana per gallon of container size: 7 gallons would yield 7 ounces; 30 gallons would yield 30 ounces; 50 gallons, 50 ounces; and so forth.

This principle also applies to plants grown in the ground. The larger the hole you dig and fill with nutritious, hi-tech soil, the bigger the root system will be, and therefore the bigger the vegetative top will be, and therefore the more flowers it will produce and the greater the yield will be.

As sheriff's deputies were cutting down one patient's garden, a deputy remarked that, after 20 plus years in law enforcement, these were the biggest marijuana plants he'd ever seen. He brought out a tape and measured the largest at more than 20 feet in diameter, with a trunk girth of 19 inches! The strain was *Northern Lights x Haze*. It was grown in a very large hole filled with hi-tech *Foxfarm* soil and nutrients. (This strain will require 10 months or longer to ripen and finish properly. It is not a good choice for northern latitudes or indoor gardens.)

The nutrients of choice are *Foxfarm Organic* liquid concentrates: *Foxfarm Grow Big* during the vegetation period, May thru July; and *Foxfarm Big Bloom* and *Tiger Bloom* during the flowering period, August thru to harvest. If you prefer dry time-release nutirents, try *Foxfarm Peace of Mind*.

If you use chemical nutrients, that's what the medicine in your pipe will taste like. Even if you flush thoroughly at the end of the season, your flowers will not have developed the rich, heady aromas and flavors that can only come from organic guano, manures, compost, kelp, so forth.

To control spider mites, thrips and other pests, use triple predator mites, 2 or 3 times during the vegetative period. *Patrol* is a product I've tried for two years, and I find it to be an effective, non-toxic substitute for predator mites.

To control gophers, prayer works as good as anything. Electric noisemakers seem to be the most effective. Make the neighborhood noisy and the gophers, who have very sensitive hearing, will leave. For awhile. Though, the gophers eventually become accustomed to the noise and move back in.

Last year I tried a method I heard about from commercial farmers in the central valley. Place a stick of Juicy Fruit gum in their burrows. Only Juicy Fruit. Other flavors won't work. In about a week, the gopher activity diminishes and ceases. In some areas, a small amount of activity might recur. Continue to place the gum in the burrows. So far, it seems to work.

If you have serious gopher issues, make wire mesh baskets to contain the roots of your plants. Or plant in containers, 30-50 gallons, sunk half way or more into the ground, for stability. Plastic barrels with holes drilled in the bottom for drainage work great. Heavy plastic grow bags, 30 gallon or 45 gallon, are available at your local hydroponics store.

NOTE: Those who are sickest and in most need of marijuana haven't the know-how, the health, the time or a suitable place to grow their own. Growing a year's supply of medical quality marijuana is something that requires some basic knowledge, a secure and sunny outdoor location and the whole year to plan, plant, raise, harvest and cure enough medicine for the coming year, not to mention the physical strength and health to do the work. A sick person cannot carry heavy bags of soil or give daily care to the garden. Caregivers **must** grow the medicine for the sick.

JJ

Age: 65. Elementary school custodian and building maintenance. Condition: arthritis, fibromyalgia, clinical depression.

JJ is another patient who asked to be instructed in growing marijuana. As an elementary school custodian, he did not make enough to afford the marijuana he needed for his arthritis and fibromyalgia. So he decided to learn how to grow it himself.

He and his wife owned a tiny home in a rural Sonoma County town. They had a vegetable garden behind the house. One day when I had stopped to leave some meds and visit awhile, JJ showed me his garden and asked if I thought he could grow some marijuana there. Marijuana is a weed. It'll grow anywhere there's sunshine. And if you farm it a little bit, with high-tech soils and nutrients, the results will astonish you!

JJ chose to grow six marijuana plants each year. And his success has been similar to Mary Jane's. JJ also grows eight-foot tall marijuana plants, along with his tomatoes and corn and squash and beans and peppers. In fact, he had to increase the height of the fence between his yard and his neighbor's yard in order to keep the marijuana from being seen.

It seems a very natural thing, to grow one's own medicinal herbs in the garden with the fruit and vegetables. Part of the pleasure of living in the country is being able to produce the things you consume. The things you grow in your garden, as we all know, taste so much better than the produce you buy at the market! And the honey, eggs or shitake mushrooms you produce not only taste better than store bought, but it is pleasurable, and satisfying to produce them. The marijuana you grow for yourself can be far better than anything you will find in the co-op. You will enjoy the growing experience. And the sunshine and exercise are therapeutic.

About ten years ago, I lived in a place where I could not grow marijuana. Not a single plant. And that was a most miserable year. I missed tending and caring for my plants. Every day. I missed their companionship. And I missed having the world-class medicine throughout the following year. Sadly, I had to smoke low-grade, imported Mexican marijuana that year. What would I have done if the co-op were not there to provide it?

I promised myself that I would never let another year go by without growing at least one marijuana plant. I begin the season by planting my garden. Including the beautiful and exotic marijuana plants. And I rise each morning to begin the day in the garden, watering, feeding, weeding. I play music in the garden on a boombox. The atmosphere it brings to the garden is welcome. And the plants love the music. That is clear from the way they respond. Their favorite album is *Crystal Waves*, by David Casper. Innersense Music, Seattle.

It has become an important part of my routine, to care for my garden and to share the produce with others, the gourmet tomatoes and squash and jalapenos and melons, and at season's end, the pumpkins and gourds and dried ears of Indian corn.

And I especially enjoy sharing the marijuana. It's a joy to cut a stem of marijuana, thick with flower buds and fall color in the leaves, to lie on the table or the bed of a patient. They can enjoy the aromas and colors, and when it is dried they can smoke it and enjoy the medicine. It is a special gift that let's someone know they are special.

In addition to growing medical marijuana for smoking, I also cook with marijuana and make hashish. The hashish is for smoking and for adding to the recipes for chemo patients and others who require stronger medicine for pain or to buffer the side effects of chemotherapy. The hash has a stronger effect on pain than the flower buds. This is very helpful to the chemo patient or the man with prostate cancer.

In essence, hashish is composed of the crystals that form on the flower buds and leaf surfaces of the marijuana. By one of several different methods, they are removed and gathered into a mass. Hashish is a concentrate.

Throughout human history, hashish has been made in one of two ways, which produce two different kinds of hashish: the dense, black hashish and the light, fluffy kef. In our lifetime, a new and more efficient method has been invented. This is the one we will discuss: a.k.a. water hash, jelly hash, or bubble hash. For those interested in further knowledge of the history and methods for making hashish, I recommend the *Great Books of Hashish*, volumes 1, 2 & 3, by Laurence Cherniak, published by And/Or Press, Berkeley, CA. Lots of great photos.

At season's end, when all the large buds have been cut and trimmed and dried and stored, I harvest the remaining leaf and small buds by stripping from the stems. Wear thin rubber garden gloves; grab a stem near its base close to the trunk of the plant; pull toward yourself, stripping all the buds and leaf from the stem. Toss this material into a brown paper bag.

Spread half of this material on screens to dry, or fill a brown paper grocery bag 1/4 full and toss several times daily to prevent mold until bone dry. About 5-7 days. This dry material is run through a wire strainer or coffee grinder, and the finished product is bagged, 4 ounces in a Ziploc sandwich bag. Place these in a turkey size oven roasting bag, twist and knot. Stored airtight, they will keep all year. Use them for making marijuana oil or butter for use in recipes.

NOTE: This material counts as part of the total weight of medical marijuana that you are permitted to possess, though it is harsh to smoke and not strong enough to use as smoking medicine. If you have enough cooking marijuana to last the year, you will probably have excessive amounts. If you have excessive amounts of cooking marijuana, it is safer to store it in a location away from the property where your garden is.

The other half of the leaf/bud mix, for hashish, should be kept fresh and pliable, not dried. I like to place the material in a bag in the freezer until I want to make the hashish. I use "Bubble Bags" (available online) to make hashish. These work fine and make some of the best hashish in the world. There was a time when hashish came only from other countries, like Lebanon, Morocco, Afghanistan, Nepal. Nepalese temple balls have long had the reputation for being the best hashish in the world. And Lebanese blond kef is unique and a rare treat.

But other countries don't have the genetics available to contemporary California growers. Nowadays, as it has happened with fine wines, California has moved ahead and now produces the best marijuana and hashish in the world.

It is a great pleasure to be able to make world class hashish and share it with people suffering with cancer and chemotherapy.

GIMLI

Age: 61. Carpenter, handyman, commercial building maintenance. Viet Nam vet, 101ˢᵗ Army Airborne, demolitions expert. Condition: kidney dialysis patient, waiting for transplant; recurring cancer/chemo/radiation patient. Homeless.

A number of my patients are homeless. Some live in parks, with little campsites hidden in bushy places. One fellow lives in a storage yard for a construction company. He has found places to hide his bedroll and personal things. He comes and goes when the office is closed. They live in cars and vans, in empty storage containers, abandoned buildings. Some live in cardboard boxes. And some live in the doorways of shops and have no shelter at all. Some last a long time. Some pass away or disappear, to be replaced by others.

They don't have phones. So I can't call to make appointments. And they don't have permanent, secure places to be. So I cannot always find them on a given day. Therefore, where the homeless are concerned, it has become my habit to stop and look for them when I pass thru their neighborhoods as I make rounds to visit or deliver meds to more stable patients, in hospitals and homes.

Gimli is one of my homeless patients. He is a short, round fellow, with a bushy sand-colored beard. He wears stocking caps and looks very much like Frodo's companion, Gimli the dwarf.

For years he has lived in a van on the street, since the loss of his wife and child in an accident. He has supported himself by finding small jobs as a handyman. He needs dialysis three times weekly. And he has cancer that recurs and must be treated on a regular basis with chemo or radiation. These treatments leave him weak and sick and bed-ridden in his van for days or even weeks. He often says to me that the marijuana is the only thing that controls his nausea and helps him get food down and keep it down, or helps him get some real sleep when he is in chemo.

He has been waiting for a kidney transplant. However, being a cancer patient and a homeless person, he knows it is unlikely that he will be given a kidney. It is more likely that a patient with better chances of survival will get a kidney before he does. His situation is not hopeful. And occasionally this leads to periods of despair and depression.

Yet, Gimli is usually an upbeat, positive person. He is quick with a handshake or a hug. He has a ready smile and enjoys doing little things for others. He'll make a swing or a tree house for a child, or fix your broken gate.

Gimli has been my patient for many years. I know the places where he likes to park his van. And I will stop to visit when I'm passing through his neighborhood. I bring him marijuana. And I also bring food when I can. He likes those Chinese noodle cups, the Styrofoam cups with freeze-dried noodles and instant broth. I will bring him a case or two of these noodle cups. He goes to the 7-11 at night for hot water to put in the cup and has a hot snack against the cold.

A few weeks ago, a misguided person complained to the authorities about Gimli's van being parked near her house. The van was towed, and Gimli was arrested on a minor loitering charge. Since he has no money and was unable to pay the fees to get his van out of impound, it was sold at auction. Now he has no home and no transportation, no tools.

While Gimli was in jail, he was beaten severely by other

prisoners.   Where is the sense in locking up a homeless cancer/dialysis patient? Who gets credit for that decision? Who eats the karma? On his release, I witnessed Gimli's face swollen, red and yellow and purple and shiny, one eye closed.

With his van gone, he had no place to sleep or shelter from the storm. He went to court on the loitering charge and received a fine of more than $1000. With no job or income, he could not pay. So the court let him work it off, a few hours a week, at $8 an hour. He averages $30 a day in wages, and he is only allowed to work three days a week. His health probably would not permit him to work more, even if the system would permit it. It costs $7 for bus fare to and from a work site, which means that he is making only $23 a day, three days a week.   At this rate, if he doesn't eat, he might pay off his fine in a year or so.

This sick, homeless man has become a legally indentured slave. He has no way out of this prison.   And if he continues to live on the street, while trying to work off this fine, the likelihood is high that he will find himself in jail again, with another fine that will take even longer to pay off.

My heart went out to the guy. This is a Viet Nam vet, Army Special Forces, living on the street with cancer and bad kidneys, being victimized by the system. The system that is supposed to be in place to help has completely failed this person.

UPDATE: As I work on this manuscript, I've just received word that tests following Gimli's most recent treatment show that the cancer is migrating, no longer responding to treatment. His situation worsens. He has informed the court that he is too ill to work and pay off the fine. He's asked to be put in jail, where he will have food, shelter and what passes for medical care. His situation is so insufferable that he feels jail is a better choice. Good luck, buddy! May the gods bless and keep you.

Living on the street is tough enough. Or dealing with cancer. Either one is hard. And doing it alone without the support of family or friends makes it harder yet. But living on the street with cancer is something most people can't even imagine. And the sad thing is that I encounter people living with this situation far too often.

MOON WOMAN

Age: 55. Buddhist nun, Native American singer. Condition: ovarian cancer survivor. Homeless.

I met Moon Woman at a recording session in San Francisco's Castro district. I was producing and taping a live performance of Tibetan music for broadcast on a Buddhist radio program. Moon Woman had come at the request of the performing artist to sing some Navajo and Zuni songs to introduce his performance. She has a beautiful voice, high and clear. Her contribution to the broadcast was most welcome.

After the session, several of us went for food at a nearby pizza & pasta shop. I learned during conversation that Moon Woman was homeless and dealing with ovarian cancer. The relationship she had been in just ended and Moon Woman had to leave. Thinking the relationship was good, she had sublet her San Rafael apartment for the next year. Having no place to go or resources of her own, she stayed with different friends for a while, sleeping on their couches.

But these arrangements were uncomfortable for her, inconvenient for friends. So she found a vacant storage locker off the garage beneath an apartment building. She put a mattress on the floor, and her own padlock on the door. Thus, she had her own private space. She was careful to go in or out very late at night or early in the morning. She showered at the local YMCA.

She was using marijuana, when she could get it. For the most part, food and marijuana were handouts from friends. She often had neither. So I volunteered to be her caregiver.

Some time before we met, she had undergone surgery for ovarian cancer. But the surgeon closed her up almost as quick as she was opened. Later, during a consultation, he told her that he had found the cancer more widespread than he'd expected, and growing in places and in ways he was not expecting or prepared for. His advice was to heal from the present surgery and try again another time. Meanwhile, perhaps other therapies might be used to shrink the cancer.

I was impressed by Moon Woman's strength, discipline and courage. She declined chemo or radiation treatments, choosing to embrace natural and alternative therapies.

Living on the street or on the couches of friends, she went through periods of planned fasting. When eating, her diet was vegetarian, very light and clean. She used Tibetan herbal medicine and acupuncture. She practiced Tai Chi Chuan and Chi Gong every day. She took daily saunas at the local YMCA. She went to a reservation in Arizona for a while, where she did sweat lodge, fasting and meditation.

After a year had passed since the first surgery, her oncologist scheduled her for surgery again. This time she was in a small private hospital in the East Bay. I went to the hospital with her. Before she went in for surgery, I promised that during her surgery I would be outside the hospital in the sunshine, walking circles around the building, offering incense and prayer, to invite the gods and make the devils sweat.

I gave the surgeon a gallon Ziploc bag with several CDs of quiet new age and Asian music. I inform my patient's surgeons that studies show that, when music is played during surgery, the percentage of successes goes up and speed of recovery is increased. I asked him to play these albums during Moon Woman's surgery. He was happy to oblige. Afterwards, some of the team commented that it was nice to have the music in the surgery, and maybe they should do it more often.

When Moon Woman was returned to her room after surgery, she was unconscious. I kept the light low and had the same soft music playing on a boombox, establishing a soothing atmosphere.

Moon Woman returned to consciousness slowly and gently. I was not aware she was awake until she spoke. An album had ended, and the room was silent. I just sat still in the quiet, listening to the soft humming sound of the hospital. Moon Woman said softly, "Play the music some more, please." So I put another album on.

She went in and out of consciousness for the next 14 hours. Each time the music stopped, if I did not replace it right away, Moon Woman would ask quietly for more music. Sometimes she asked for more pain meds.

After her recovery, when she spoke to her surgeon, she learned something astonishing. He had found no cancer. That's right: zero cancer. There was scar tissue everywhere, but no cancer. So he had closed her up again, this time with good findings.

We are guessing that the year of fasting, meditation, sweat lodge, Qi Gong and herbal medicine had worked a miracle. However you explain it, Moon Woman is now cancer free. She is a strong and highly disciplined woman. It has been a memorable experience for me to watch and support this medicine woman dealing with her own cancer. Happy Trails, Moon!

MUSIC THERAPY

At MD Andersen Cancer Center, the largest cancer research hospital in the United States, much has been done to study sound and its effect on cancer. Studies show that when music is played during surgery, the percentage of successes goes up and speed of recovery is increased. Knowing this, it seems to me the better part of wisdom to give ourselves every possible advantage.

Therefore, for patients' surgeries, and for atmosphere in the hospital rooms, I urge the use of music. I use a collection of new age albums, offering the surgeon a gallon Ziploc bag full of CDs, and a note informing him of the studies at Andersen and requesting that he play the albums during the surgery. These requests have never been denied.

Most of these albums are by spiritual artists, who have studied the science as well as the art of sound. They are conscious of using sound to stroke, soothe and re-tune the human organism and its nervous system, to refresh and relax the listener. Almost all of this music is based on a drone.

THEORY: All mass, all physical forms, including our bodies, are composed of atoms, which are patterns of energy, frequencies. As we expose ourselves to the vibrations and paces of modern life, through the principle of "sympathetic vibration," our electrical patterns become disturbed, de-tuned from our natural frequency and re-tuned to the frequencies of machines, power tools, traffic, electric and magnetic fields, and countless other frequencies and vibrational influences in the world around us. Being out of tune or balance leads to physical and psychological health issues. To re-tune helps us to avoid or to heal from sickness.

EXAMPLE: We all know that a vibrating tuning fork, when placed next to an unmoving tuning fork, will cause the unmoving fork to begin to vibrate: "sympathetic vibration." We also know that we are each affected by sound in tangible ways. For example, when fingernails are dragged across a blackboard, the hair on the backs of our necks stands up and we get goose pimples. The sound of screeching tires close by will cause our adrenaline to peak. We are two-legged tuning forks constantly being affected by the sounds going on all around us. Some musicians, aware of this phenomenon, consciously use the sound of their music to effect beneficial change in the human organism.

The Tibetans have been on to this for millennia. Researchers at Andersen are catching up, though.

It is my personal take on the subject that music by these spiritual artists is probably more effective than arbitrary selections. Therefore, these are the albums I've selected for use with patients. Also, I enjoy the soothing, relaxing atmosphere these albums bring to the hospital environment. I spend a good bit of time in hospitals with patients. The music is relaxing for me, as well.

But in the studies, I believe that any and all kinds of music were used effectively, from classical to rock. Bottom line: it seems more important that you have music during surgery than what kind of music it is.

Here are the core albums in the collection. Each is a quiet album and has been chosen for use in the hospital environment:

Crystal Waves. David Casper.*
Tantra-La. David Casper.
Earthsight. David Casper.
Unwrap. David Casper.

Raku. Philip Davidoff.**
Santosh. Philip Davidoff.
Secrets of the Jade. Philip Davidoff.
Bamboo. Philip Davidoff.
Lush Mechanique. Jami Sieber.
Golden Bowls of Compassion. Karma Moffett.
Way to Katmandu. Karma Moffett.
Red Wind. Carlos Nakai.
Island of Bows. Carlos Nakai & Wind Traveling Band.
Winds of Devotion. Lama Nawang Khechog & Carlos Nakai.
I Will Not Be Sad in this World. Djivan Gasparyan.
Aerial Boundaries. Michael Hedges.
Bach: Complete Sonatas and Partitas for Solo Violin. Arthur Grumiaux.

Any quiet music selections are suitable. I choose many new age artists and Asian music albums. Offerings from **Windham Hill** are good choices. Relax and let the sound of the music wash over you and through you, caress and soothe you, as the cool waters of a mountain stream or a gentle breeze.

*David Casper's albums are particularly strong medicine. *Crystal Waves* is the best of all musical medicine. But they can be hard to find. If you are going to use music in the surgery or hospital, David Casper's music should be at the top of your playlist. If his albums are not available from your retailer, order from Innersense Music, Box 30714, Seattle, WA 98103. All albums in his "Quiet Music" series are good choices.

**Phillip Davidoff's albums may also be hard to find. They are among the best you'll find for use in quiet environments. And the search will be worthwhile.

SUICIDE BOB

Age: 62. School administrator, Saigon, Viet Nam. Condition: terminal cancer.

This chapter on suicide has been included because it's an issue that some terminal cancer patients think about and talk about. Other caregivers may have to confront this issue. There are things I've learned that could be helpful.

On learning he had terminal cancer, Bob elected to return from Saigon to San Francisco, a city he loved and a place where he had friends who might offer support as he approached his end.

One friend did indeed offer a comfortable downtown apartment for Bob's use. As he grew weaker and his health declined, he became deeply concerned about reaching a point where he would become helpless and completely dependent on others. He was morbidly afraid of being institutionalized and kept alive artificially. It became important to him to have a means of ending his own life, on his own terms, with dignity.

Being too sick and weak to seek information himself, he asked me to seek information for him that would help him safely and comfortably end his own life. I made it clear that as a Buddhist I would not assist a suicide, but simply help him gather information for his own peace of mind. I include this discussion here, because of what I found as I sought information for Bob.

As I began the search, I went to Hemlock Society websites, and assisted suicide webs, and so on. There are dozens, maybe hundreds. Taking note of their recommendations, I then began to discuss the issue with several doctors. They discouraged the use of the methods recommended on these websites, saying that at least two of those methods--an overdose of helium or an overdose of pills--have as much chance of leaving the patient brain dead or in a coma as deceased. The patient may not get enough helium to do the job. Or they might not take enough pills. Or the body may regurgitate some of the pills.

One of the doctors recommended a different method, the use of multiple (10) skin patches of a powerful anesthetic/pain killer, *Fentenyl*. These can be measured precisely and cannot be regurgitated. Others I questioned about this method agreed. This seemed to be the only method that all the doctors approved of.

After sharing my findings with the patient, he seemed relieved. He said he was already using the recommended patches for pain, and he could save enough to keep on hand for his time of need.

At that time, Bob had only three months to live. That was more than four years ago. He has not ended his own life. Though, he has lived with far greater peace of mind knowing he could do so if he wished

NOTE: I believe that Bob's marijuana use kept his spirits elevated and prevented morbid thoughts and was a strong deterrent from a need for suicide.

Bob has made the request that when it is time for him to pass on, he would like to be taken outdoors so that he can see the sky, hear the birds and smell the flowers. He asks that his ashes be scattered in a flower garden in Golden Gate Park.

MEDIA

I have seen many news stories and discussions about medical marijuana, in newspapers, magazines, on the evening news, and even on the network news magazines, like *60 Minutes*. But I have never seen a thorough, objective story about medical marijuana by a serious mainstream journalist.

Always, the reporter offers a regurgitated version of outdated, false and misleading information handed out by the feds for public consumption. And the feds are very clever with lies and misinformation. Never have I seen this information questioned or challenged. Then there are a few images of ragged potheads in front of the local co-op, and interviews with neighbors of the local co-op. That's about it. No imagination. No work ethic. Just re-write the fed's news release.

In one case, the feds cut down a co-operative medical garden supplying 75 patients. The cut and dried flowers from the immature garden might have amounted to 50 or 75 pounds. The feds took the marijuana to a feed and grain store to weigh it on the feed scale. They weighed entire un-dried plants, including leaves, stems, trunks, roots and even some soil. In their news release, they reported that they had seized a ton of marijuana! The local reporter printed that in his article, without a question. Shame on lying, unethical federal agents, and shame on lazy journalists.

Not once have I seen a story that includes interviews with cancer patients, loved ones, doctors, nurses, oncologists, caregivers, researchers, lawyers, judges, lawmakers, farmers.

I have not heard hard questions such as: Why does our government put pressure on Canada, Mexico, the Netherlands, the UN and the world community to block legalization, research and medical use? Why is the government spending so much money and effort to block legalization, research and medical use in the U.S.? Why does the government continue to spend money and effort to arrest patients and caregivers in the growing number of states that have legalized medical marijuana? Why does the government spend so much effort to intimidate, harm and make examples of doctors who advocate medical marijuana and actively support patients, like Tod Mikuriya and Marion Molly Fry? Why did the lawyer arguing the Angel Raich medical marijuana case before the U.S. Supreme Court make the accurate prediction that his case was solid enough to win in any court in the world except the U.S. Supreme Court?

AUTHOR'S PERSONAL STATEMENT

Age: 66. Buddhist priest, poet, author; Protestant minister; Viet Nam vet. Condition: heart patient; severe arthritus; fibromyalgia; chronic liver, kidneys, stomach, prostate.

I live with constant pain throughout the body, and zero strength or energy. For almost 20 years I've found relief by smoking marijuana 24/7, without contraindication. I smoke a briar or deer antler pipe, or I roll joints. I do not use vaporizers. I eat a marijuana-medicated confection with every meal or snack, 5-6 times daily.

The comic, Rodney Dangerfield, stated in an interview that for his health issues, for 30 years, he also smoked 24/7, without contraindication, and ate marijuana brownies with every meal.

Without the marijuana, I am in such physical discomfort that I can focus on nothing else. I have no appetite, and I cannot sleep, unless utterly exhausted. My energy is so low that I cannot sit or stand very long and spend my time in bed or on the couch, while the days pass and life ebbs away. I'm constantly depressed. Without marijuana, I am a useless vegetable.

With the marijuana, the pain is reduced substantially and is bearable, the inflammation is reduced, and my energy is sufficient to pursue some kind of useful life. I can avoid the depression and negative mindset.

The quality of the marijuana I use is important. The better the quality, the more active I can be, the more work I get done, and the quality of my work is improved. Use the best you can get.

I do not choose to spend the rest of my days on the couch in a stupor. The choice is a no-brainer. If I am to live a useful, productive life, marijuana must be a part of my protocol. The government notwithstanding, the choice is a matter of survival and quality of life. The government has no business making the choice for me, deciding how good I should feel. Marijuana works! And I'm going to use it!

To the oncologists who prescribe *Marinol* and advocate medical marijuana: Please advise your cancer patients to begin marijuana therapy early in their treatment, or refer them to a professional caregiver early on. The marijuana will be most helpful when it is part of the protocol from the beginning. Do not wait, as some oncologists do, until options are gone and marijuana is a last resort, to comfort the hopeless and dying patient. If you wish to do the best for your patients, encourage them to get with marijuana before surgery, to use it in the hospital during recovery, and throughout the chemotherapy. In my experience, it is also wise for the cancer patient to continue marijuana therapy as a permanent part of the protocol, to aid in the prevention of a recurrence.

Many chemo patients do not like the effects of *Marinol*. For these, counsel them to try marijuana instead. It is so much better, more effective, and less costly.

And finally, do not hesitate to recommend marijuana for children and young people. I observe that the young are able to embrace marijuana therapy more quickly and completely and to benefit more from its use than older patients. For the very young, smoking is not necessary. An oral dose with meals is fine.

LEGAL INFORMATION

My lawyer is at the Pier 5 Law Offices in San Francisco. He is a highly knowledgeable and experienced marijuana lawyer. There are many experienced marijuana lawyers at Pier 5 Law Offices. They are good people. If you are going to be a caregiver, consult with an experienced marijuana lawyer.

My lawyer has reviewed the following material and deems it to be sound information:

Many arrests for marijuana possession are due to traffic violations and noise complaints.

Travel safely: Do not smoke while driving. Keep the passenger area free of the odor of smoke or of green marijuana. Don't leave roaches in your ashtray. If you carry marijuana in your vehicle, make sure your vehicle is up to code, registration is current and that your marijuana and related items are stored airtight and concealed, preferably locked in your trunk.

Be a good neighbor: Loud music and domestic disputes can bring law enforcement to your home. Remember the brothers Green!

Be discreet: Try not to smoke where others can see you or smell the aroma. Never leave marijuana or related items in plain view, at home or in your vehicle. And don't let it get taller than your corn! Remember the brothers Green!

Do not be too generous: Giving marijuana to anyone, even to a homeless person, unless you are their caregiver, is a crime. If the amount is an ounce or less and no money changes hands, it is a minor offense. But know your recipient; he may be a narc!

Do not consent to a search: If law enforcement comes to your home or stops your vehicle, do not consent to a search. If an officer says "Do you mind if I look in your purse, bag, home or car?" You respond by saying "I do not consent to a search." If the officer says "Why not? Are you hiding something?" You should respond "I don't have time. I do not consent to a search."

Do not let an officer into your home without a warrant or court order: If an officer knocks on your door, step outside and close the door behind you. Never leave a door or garden gate open in the presence of an officer. Police officers often have poor manners and will interpret open doors or gates as an invitation.

Exercise your rights: 1) If you are stopped or questioned by an officer, ask if you are being detained. If not, walk away.

2) If you are being detained, ask why. Ask the officer to cite the law, and write down or remember what they say.

3) If you are being arrested, say "I choose to remain silent, until I talk to my lawyer."

Then call Pier 5 Law Office: 415 986-5591.

If you are going to be a caregiver, see an experienced marijuana lawyer, to be sure you are well informed and have the right paperwork.

If you have problems, questions or concerns about your rights, or if you wish to have support in the hospital, speak with one of the lawyers at Pier 5 Law Offices. 415 986-5591

RESOURCES

MARIJUANA FRIENDLY LAWYERS

Omar Figueroa
James Jason Clark
Shari Lynn Greenberger
Nedra Ruiz
Sara Zalkin
Laurence Jeffrey Lichter
Randolph E. Daar
J. Tony Serra

Pier 5 Law Offices
506 Broadway
San Francisco, CA 94133
415 986-5591

MARIJUANA FRIENDLY PHYSICIANS

Tod Mikuriya, M.D.
510 548-1188

Marian Molly Fry, M.D.
530 272-9901   530 823-0446

Eugene Schoenfeld, M.D.
415 331-6832

Frank Lucido, M.D.
415 457-6771

Hanya Barth, M.D.
415 255-1200

Jeffrey Hergenrather, M.D.
707 824-8444

Philip A. Denny, M.D.
916 660-9205

David Bearman, M.D.
805 899-1700

Claudia Jensen, M.D.
805 648-5683

Helen Nunberg, M.D.
831 466-9999

William Eidelman, M.D.
323 463-3295

MARIJUANA AUTHORITY & EXPERT WITNESS

Chris Conrad
510 215-TEAM

CO-OPS

Berkeley Patients Group
510 540-6013

Harborside Health Center
510 533-0147

NOTE: There are hundreds of co-ops, and they are easy to find. The greatest concentration is found throughout San Francisco, Berkeley and downtown Oakland, in the "Oaksterdam" neighborhood (17th & Telegraph). The feds close some of them, and they re-open down the street. And there are different degrees of cleanliness in force.

Berkeley Patient's Group and **Harborside Health Center** are two of the best co-ops in the Bay Area. Their reputation for offering high quality, mold-free marijuana and vigorous, healthy, disease and pest free plants has no equal I'm aware of. Both offer seed. Both offer theraputic massage and/or acupuncture. **Harborside Health Center** offers freebies for low income or hospice patients. **Berkeley Patient's Group** offers freebies and weekly visits for hospice patients.

GROWING SUPPLIES

All the growing supplies mentioned in the narrative will be found at or can be ordered from your local hydroponics shop or nursery. **Berkeley's Secret Garden** hydroponics supply is well stocked and clean. The staff is friendly and knowledgeable. **Harmony Farm Supply and Nursery** in Sebastopol is also well stocked and clean. The staff is friendly and knowledgeable. Lab work and crop specific soil analysis are available.

REFERENCE

Marijuana Medical Handbook: A Guide to Therapeutic Use. Tod Mikuriya, M.D.

Hemp for Health. Chris Conrad.

Marijuana Horticulture: The Indoor/Outdoor Medical Grower's Bible. Jorge Cervantes.

Marijuana Cooking: Good Medicine Made Easy. Cameron & Green.

Great Books of Hashish. Vol. 1, 2 & 3. Laurence Cherniak.

The Ruba'iyat of Omar Khayyam. Peter Avery & John Heath-Stubbs.

Foxfarmfertilizer.com

Yahooka.com